THE VICTORIA & ALBERT MUSEUM
INDIAN ANIMALS DAYBOOK

PUBLISHED BY
The Victoria & Albert Museum, 1990
IN ASSOCIATION WITH
The World Wide Fund for Nature

COPYRIGHT
The Trustees of the Victoria & Albert Museum

TIGER PATTERN ON COVER
REPRODUCED BY PERMISSION OF
Tiger Rugs of Tibet Fund

PICTURE RESEARCH
Jennifer Blain & Lesley Burton

BOOK DESIGNED BY
Carroll, Dempsey & Thirkell Ltd

PRINTED AND BOUND IN ITALY BY
Arnoldo Mondadori Editore, Verona
ISBN 1 85177 008 9

THE
VICTORIA & ALBERT MUSEUM

Indian Animals Daybook

IN ASSOCIATION WITH
The World Wide Fund for Nature

INTRODUCTION BY
Mike Carwardine

CAPTIONS BY
Professor Graham Chapman,
Rosemary Crill,
Jennifer Blain
&
World Wide Fund for Nature

VICTORIA & ALBERT MUSEUM

It is possible to generalise about many places, but not India. This vast country of mountains and rivers, deserts and forests, small rural villages and burgeoning cities is a land of contrasts. The seventh largest country in the world, it stretches 2,000 miles from the Himalayas in the north to the tropical Indian Ocean in the south; and 1,850 miles from Pakistan in the west to Burma in the east. India is exhilarating, mystical, bewildering and, at times, infuriating. But it is never boring.

During the country's tumultuous 4,000-year history, countless invaders and settlers have left their mark, so there is no such thing as a 'typical' Indian. If ever a government had difficulty communicating its policies to its people, it is that of India: 1,600 different languages and dialects, written in 13 alphabets, are used by the profusion of ethnic and cultural groups.

India is very poor. Although it has a well-developed industrial structure, most of its people live traditional lives seemingly untouched by the 20th century.

More than 80 per cent of the people are Hindu, which is why most Indians have a great respect for wildlife. Unlike the Western Imperial tradition, which has taught that animals are in the world solely for the benefit of people, Hinduism teaches that all life is sacred. Paintings, sculptures and street theatre drawing on the rich Indian culture, tell a fascinating story of the traditional empathy the Indian people have with the animal world.

India has a rich and varied wildlife. It is home to more than 13,000 flowering plants, 50,000 insects, 560 reptiles and amphibians, some 1,200 species of birds and more than 500 different mammals. Many of them are found nowhere else in the world. It is possible to see 200 different birds in a day, although most of the 'big game' animals are more secretive, preferring to remain well hidden in the safety of the forests. However, with a little patience and luck it is possible to see tigers, lions, leopards, rhinos, crocodiles and many other exciting creatures.

The best way to search for wildlife in India is on the back of one of its well-trained elephants which are a familiar sight in many parts of the country. Valued as beasts of burden for hundreds of years, these intelligent and gentle animals are used for religious ceremonies and in forestry and tourism.

The traditional Hindu respect for life is one of the main reasons why so much of India's rich wildlife has survived in such a highly populated country. There are people everywhere: 800 million of them

lining the roads, balanced on the tops of trains, clinging to the sides of buses and bulging out of rickshaws. Their numbers continue to grow, despite an active family planning programme, by an astonishing 1.3 million every month. It already has the second largest population in the world (after China), and it is expected to pass the one billion mark by the end of the century. Yet even in the busiest towns and villages, there are hundreds of different animals to be seen alongside the famous sacred cows, which are free to wander wherever they like.

Despite this reverence for nature, in the last three centuries habitats have been lost and many species have been driven to the brink of extinction. Early paintings and crafts illustrate how the wildlife and landscapes of India have changed over the years. Forests that were painted centuries ago have since turned to desert, and animals that were once common and widespread have virtually disappeared. It is interesting, and rather disquieting, to see how many pictures painted during colonial times were based on hunting parties.

Big cats were the hunters' favourite quarry. Perhaps they believed that the lion, cheetah and tiger populations were inexhaustible, or probably they didn't care, but they slaughtered them, along with many other animals, on an unprecedented scale.

The Asiatic lion, for example, was once very common, especially in the north. But so many were killed that, by 1940, they were extinct everywhere except one place: the Gir Forest in the state of Gujarat. Nowadays, Gir has a remnant population of about 200 lions – the only ones surviving outside Africa.

Cheetahs were also killed, or captured, in large numbers. Famous for their speed, they made popular pets and were often trained to hunt chinkara gazelles and blackbuck: the Emperor Akbar used to maintain a contingent of 1,000 of the animals for his hunting expeditions. India's last wild cheetah was shot in 1947.

But tiger hunting was the most popular sport. At the turn of the century, there were some 40,000 tigers in India, living wherever there was water, prey and somewhere to hide. So many were killed in the early 1900s that several Indian maharajahs (who frequently organised enormous hunting parties in honour of visiting dignitaries) claimed to have shot more than 1,000 each. As the news spread that the tiger was heading for extinction, hunters from all over the world began to arrive in India to 'bag one' before it was too late. By 1972, there were only 1,827 of the animals left.

More than twenty years after the British left India, a new revival of conservation-mindedness came just in time, with the strong personal support of the Prime Minister, the late Indira Gandhi. Concerned that the tiger was in serious trouble, she banned tiger hunting in 1970. Three years later, Project Tiger was launched by the Indian Government, with assistance from the World Wide Fund for Nature, and a network of closely-guarded reserves was established in a desperate bid to save the tiger from extinction.

Project Tiger has been highly successful. There are now as many as 4,330 tigers in India (although more pessimistic estimates put the figure at nearer 3,000) and the population is healthier than it has been for a long time.

But the tiger's future is still precarious. Under pressure from the expanding human population, the forest habitat it shares with so many other species is disappearing at an alarming rate.

Forests were greatly valued and carefully looked after in ancient India. They were used, but on a small scale, and little permanent harm was done. But vast areas were cleared by the invaders and colonisers. Valuable trees such as sandalwood, rosewood and ebony were felled for commercial purposes and others were destroyed to make room for cash crops such as tea and cotton. The plundering continues today, with India's insatiable demand for timber and paper.

The people who suffer the most from this destruction are the poor; and they add to the pressures on the forests by collecting firewood for cooking, fodder for their livestock and forest plants for their food and traditional medicines.

Both rural and urban households in India are heavily dependent on firewood. But it is already very scarce in some areas and, within the next decade, demand for it is expected to more than double. Whenever firewood in unavailable, people are forced to burn cow dung or crop residues instead (which should be left to fertilise the soil) or they turn their attentions to the rich forests of national parks and reserves.

Livestock is also a problem. India already supports an incredible 1.5 billion domestic grazing animals and its forests are among the most heavily grazed anywhere in the world. Many of the buffaloes, cattle, sheep and goats are taken into parks and reserves to feed: some 50,000 rely for fodder on Ranthambhore tiger reserve, for example.

Forests are also important to people and wildlife indirectly – for water conservation. They hold moisture like a sponge, releasing it

slowly and steadily. This is very important in a monsoon climate, like India's, with a highly uneven rainfall throughout the year. Without forests, there are more floods in the rainy season and longer droughts in the dry season; and heavy rain washes the unprotected topsoil into lakes and rivers, where it silts up the riverbeds and chokes hydro-electric dams.

The protection of India's remnant forests has become one of the most urgent environmental issues facing the country today. Already, they cover as little as eight per cent of the country – and the more they deteriorate, the more the wildlife and people will suffer.

But there is more to saving India's forests than setting aside protected areas with strict rules and regulations designed to keep people out. In a country with such a high level of poverty, it would be irresponsible and impractical, to put the welfare of the forests themselves (and of their wildlife) before people. Conservation must take a pragmatic approach.

There are many experts in India who are fully aware of the need to bring people and their environment into harmony. They have published an extraordinary document called 'The State of India's Environment' (more commonly known as 'The Citizen's Report') which is an invaluable commentary on the current environmental issues facing India. It is, above all, a powerful plea to encourage wider use of the country's natural resources, for the benefit of the poor as well as for the rich.

Rural women are often keen supporters of such conservation activities. In urban India, many women have top jobs running businesses, managing hotel chains or editing magazines. But in rural areas, where three-quarters of Indian women live, the society is still male-dominated. These women have to spend most of the day, seven days a week, collecting essential resources such as firewood, fodder and water from the surrounding countryside. The economy of rural India depends on their unpaid labour. But as the environment is degraded, the women have to spend longer and longer searching for these household essentials, so they are often keen to support any conservation activity that will make their lives a little easier.

There are local women's groups which organise their own reforestation programmes. One of these is the Chipko Movement, which was established in 1974 after the women of Reni, in northern India, took direct action to stop the commercial felling of a large woodland near their village. It was a simple protest – they threatened to hug the trees if the lumberjacks came too close – but the campaign was highly successful. Now the women organise tree planting schemes in other villages in the region.

'Recreating nature' in this way has been successful in some areas – but many national tree planting schemes tend to involve wealthy farmers and foresters, more interested in making a profit for themselves than meeting the needs of the poor. Tree planting still fails to keep up with rates of deforestation and more appropriate schemes need to be developed on a much larger scale.

Forest conservation in the long-term must involve the active participation of local communities. Indeed, leading experts in India believe that integrating nature conservation with the needs of rural people is the *only* solution to many of the country's environmental problems.

Project Tiger is now developing along these lines. An excellent example is the Palamau Tiger Reserve, in Bihar, which successfully combines wildlife conservation with the sustainable use of natural resources. The reserve has a fully protected core area, for wildlife, and a large buffer zone for local villagers. The people graze their cattle and harvest forest products such as bamboo, grass, wood and wild fruit and are employed in the construction of patrol roads, guard huts and firebreaks. As a result, their own future is more secure and they have a vested interest in the future of the reserve.

This kind of approach is in stark contrast to most conservation activities in the industrialised world, where development has resulted in a growing alienation between people and nature. India is keen to avoid making the same mistake, which means modifying the development process itself.

One of the country's leading environmentalists, Anil Agarwal, captured the essence of this exciting approach when he said, 'It is not enough to preserve biological diversity just in those areas of India where the flora and fauna are rich and diverse. Instead, we must ensure that it is preserved, or created, in every village ecosystem.'

This is the challenge facing India today.

Mark Carwardine

This lady with her pet is typical of the romanticised female portraits painted in Bikaner in western India in the 18th century. Bikaner painting is unusually refined for such a remote Rajput kingdom, far from the Mughal court: artists trained in the Mughal style had settled in Bikaner by the mid 17th century, and entire families of these Muslim artists produced high quality work right up to the 19th century. The empty background is typical of these portraits, and gives an idea of the vast sandy expanses of the Thar desert in which Bikaner is situated. Painted by Ahmed, son of Muhammad in 1731. IS. 337.–1951

JANUARY

Illustration to the musical mode *Kakhuba ragini*. Each of the
main *ragas* in Indian music is intended to create a certain mood and effect, and have
become associated with particular seasons and times of the day or night. In northern
India, they were depicted in paintings expressing the mood of the *raga*: *Kakhuba* is
traditionally shown as a lovelorn lady wandering in the forest, alone but for peacocks
and wild animals. Here there are deer, hares and monkeys. Two Banyan trees
with their roots hanging are in the foreground. Servants cluster around
the water tank and the Palace is behind. This illustration was
painted in Lucknow in about 1760. IS. 158 – 1952

JANUARY

JANUARY

Illustration to the popular but melancholy musical mode *Todi ragini*.
The landscape is reduced to a decorative framework for the scene of the lady sadly
playing the *vina* to four black buck. Below, the waves of a stream are reduced to a
linear pattern, and above is a symbolically dark, clouded sky. Along the top is
a purely decorative row of naturalistic flowers in a style derived from
Mughal paintings. Probably painted at the Rajput court of
Amber in the 18th century. IS. 75 – 1952

Overleaf. This scene of a tree full of crows is probably an illustration
to a fable, perhaps the popular Sanskrit work the *Panchatantra* ('Five Books').
This painting, done in Bundi, Rajasthan, in the mid-19th century, shows the
common house crow, which is a scavenger and therefore unclean. IS. 247 – 1952

The use of numerous animals and servants to emphasise personal wealth
and power is a concept also familiar in western art, yet this is an illustration in the
history of the Emperor Akbar's reign, the *Akbarnama*. It was painted by Abu'l-Fazl
around 1590 and is a good example of the recording nature of Mughal art. The
Emperor is hunting black buck with trained cheetahs, and his men carry weapons
wrapped in a red cloth. A wide variety of plants can also be seen. This wilderness
was preserved to encourage game to live and breed in hunting areas, a number
of which have formed the basis of modern National Parks
and reserves. IS. 2 – 1896 92/117

FEBRUARY

FEBRUARY

This watercolour was painted in Delhi in 1827 by an Indian artist for a
European patron. It shows the Mughal Emperor Akbar II who ruled from 1806 – 1837.
The elephant which carries him is fully caparisoned, its trunk, face and ears have
been gorgeously painted; its tusks, sawn short through the solid ivory,
are banded in gold and it wears lavish cloths. IS. 59 – 1964

Akbar II, King of Delhi
And the Prince Mirza Selim, his Son.

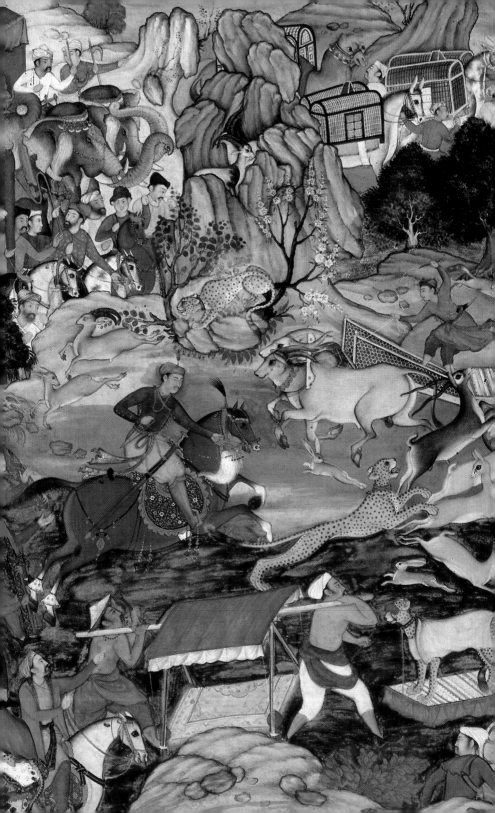

Akbar is shown hunting near Agra. He eventually became sickened
by slaughtering animals and ordered it to stop, though continuing to hunt himself.
This painting is by the artists Basawar and Dharmdas. It was customary for more
than one artist to work on a single painting, one concentrating on the outline,
another on the detail. Here, empty cages, used for transporting the cheetahs,
as well as a camel and elephants can be seen at the top of the scene.
A blindfolded cheetah standing on a covered platform
can also be seen. IS. 2 – 1896 24/117

MARCH

MARCH

Watercolour of a blossom-headed parakeet by Hulas Lal, Patna,
about 1850, probably painted from a dead or tame specimen to allow for accuracy.
Common, brightly plumaged birds are often painted, unlike common animals
or even people unless they appear as part of a royal retinue where they
serve to emphasise the ruler's power. IS. 96 – 1949

Overleaf. "Shaul goat from Boutan". Protected by its long staple wool,
these goats can live high in the Himalayas. Although known in Europe for
producing the wool for Kashmir shawls, the goats were found over a much wider
area. Domestic goats, in large numbers, can do a great deal of damage by
rearing up on their hind legs and browsing the new growth. Painted by
Zayn-al Din in Calcutta in 1778. IS. 51 – 1963

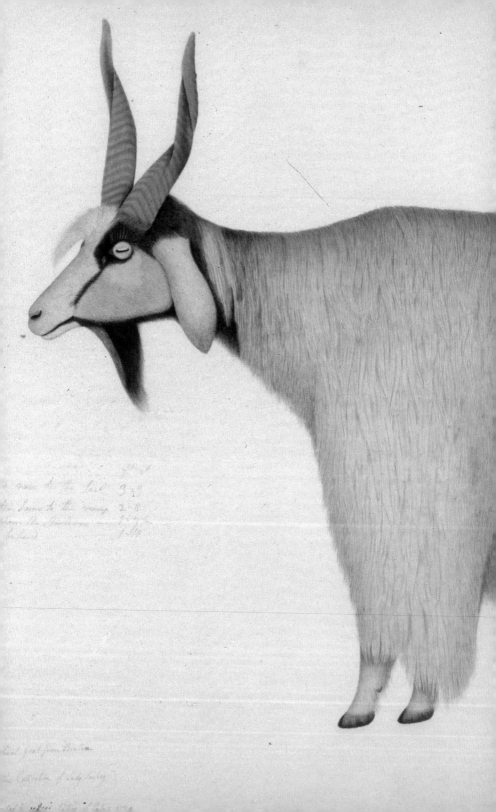

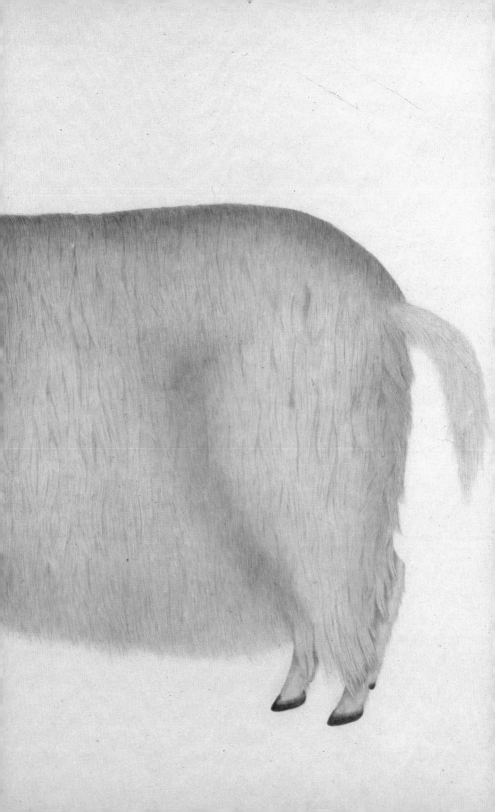

Radha and Krishna adored by cows. The god Krishna fell in love with
the mortal cow-keeper Radha, and many paintings illustrate episodes from their
courtship and lovemaking. Cows are particularly associated with Krishna, and are
revered throughout India as providers both of milk and of bullocks to work in the
fields. All the products of the cow are considered ritually pure – cow dung for
example, is beaten and polished as a floor-covering, and milk is habitually offered to
the gods in worship. Cow worship entered the Hindu religion in the early centuries
AD, and by the fourth century it was an offence to kill a cow. Mahatma Gandhi
considered that 'Cow protection is the gift of Hinduism to the world.' This
scene was painted in the Punjab Hills in the mid 19th century. IS. 230 – 1950

APRIL

APRIL

This watercolour of a lady with two cranes is probably an
illustration to a *raga* or musical mode. It was painted in Bundi, Rajasthan
in the mid 18th century. The birds are widespread
throughout the country. IS. 169 – 1955

A princess and three ladies on horseback hunting black buck.
Women are quite often shown hunting or playing polo, and there was also a
convention of showing a male rule accompanied by an all-female 'army' in
procession. Although such scenes superficially imply a certain amount of freedom
or equality for women, all the women shown in the paintings would be members of
the *zenana* (women's quarters) of the royal palace, and their activities would revolve
around those of the male ruler. Here the buck has apparently been stuck with
daggers and is falling to the ground, and male retainers armed with clubs accompany
the ladies, presumably to finish off the prey. The landscape is a stereotyped Persian-
influenced scene with dramatic pink rocks which probably represent the tors
in the Deccan where the picture was painted in about 1730. IS. 293 – 1955

MAY

MAY

A lady with a bird, possibly a crossbill, on her hand.
The painting is highly stylised and the artist has treated the bird as if it is merely an
accoutrement like a jewel or a flower to enhance the woman's beauty. It was
painted at Hyderabad in the Deccan in the mid 18th century. IS. 210 – 1951

Overleaf. This print of a painting by Captain Richard Barron
is of Ootacamund in the Niligri Hills, Tamil Nadu. He was probably an amateur
artist who, like many visitors, painted for their own amusement and record before
and after the invention of photography. The deforestation of the mountain behind
Ootacamund can be clearly seen. Although there were earlier reforestation
schemes, this was to become the site of some of the first British efforts
at forest preservation in India. IM. 119 – 1918

An ascetic outside his hut. Hindu *yogis, sandhus,* and *sannyasis* –
ascetics and wandering mindicants – are often shown clad in animal skins as a mark
of their poverty and other-worldliness. The god Shiva, the 'great ascetic', similarly
wears skins, and is often depicted under a canopy made of an elephant's hide.
This hermit has befriended a tiger, recalling paintings in the Christian
tradition of St Jerome and his tame lion. Murshidabad,
Bengal, 18th century. D. 1192 – 1903

JUNE

JUNE

JUNE

A European hunting tiger in the Terai – the grassland plains on the
borders of Nepal. Grasses can grow to five metres and become so thick that it is
difficult even for elephants to penetrate. Plagued with malaria and home to wild
animals, the area was little settled in the past and is now the home of numerous
sanctuaries, particularly for tigers. The artist shows the fragile nature of the ecology
in an area where the forest and grassland can easily be degraded and destroyed by
inappropriate human use. The painting was done in Patiala in the Punjab
in 1892 and is in the style of the late Delhi school. IS. 60 – 1968

J U L Y

Because they were regarded as unclean animals boar were not hunted in
India before the British settlement. Here the haloed Maharana Jawan Singh of
Mewar in Udaipur is shown hunting dressed in green, probably during a spring
festival. In the hilly background, which is green and fertile, the *rana* again appears:
it was common practice, especially in larger Mewar paintings, to show
several episodes in the same painting. 1835. IS. 557 – 1952

JULY

JULY

Attacked by a phoenix the mythical bird Anka is grasping elephants
in its claws. The motif is widely found in, for example, wood inlay and carpets.
The lower half of the painting shows how elephants live as a family group, here
clustered round a water hole in arid country. They would probably be migrating in
search of fodder. Mughal, mid 17th century. IM. 155 – 1914

Overleaf. Krishna playing to the milkmaids and their cows,
Jaipur, Rajasthan, c.1840. Krishna's name and other epithets associated with him
mean 'dark and beautiful', hence he is depicted with blue skin. Some cowherd
communities in India today believe they are descended from Krishna and
carry wooden flutes in his memory. IS. 109 – 1951

عمل استاد منصور

AUGUST

A Himalayan *chir* pheasant, painted around 1620 by the
Mughal court artist Mansur. He specialised in finely detailed studies of birds,
animals and flowers, and the Emperor Jahangir would frequently order him to
record on paper any interesting or beautiful creature that was presented to him.
Many of these paintings would then be mounted within elaborate borders and
bound into albums for the Emperor to browse through. The *chir* pheasant
is now considered a threatened species. It inhabits open conifer and deciduous
forest interspersed with steep grassy slopes. IM. 136 – 1921

AUGUST

AUGUST

AUGUST

Raja Guman Singh of Kota, Rajasthan, shooting lion.
This very powerful picture dated 1778 achieves great tension between the lion
at bay and the taut bows. Kota lay near luxuriant bamboo forest in the mid
18th century but today its acreage is much reduced. Although the area of wild
bamboo forests has declined in recent decades, cultivated clumps are often found
near villages. Kota artists were renowned for their powerful images of wild animals,
though, as throughout the images in this book, the animals are normally seen in
association with man, either in his service or as his prey. IM. 136 – 1927

In this illustration to the musical mode *Asavari Ragini*, a lady,
dressed in a peacock feather skirt charms the snakes out of the trees. The forest
clearing and dark sky are symbolic of her pining as she awaits her lover.
Done in Chawand in southern Rajasthan in 1605, this is one of the
earliest *ragamala* series known. IS. 38 – 1953

SEPTEMBER

SEPTEMBER

Radha awaiting Krishna, Kangra, 1820-25. This typically romantic
late Kangra painting shows Radha in a clearing hemmed in by forest trees.
A crescent moon barely lights the scene. Forest animals and peacocks and
ducks on the water watch her vigil. IM. 157 – 1914

Overleaf. This late 18th century painting from Rahjasthan could be an
illustration to a folk tale. A tiger hunts by sight. Both its eyesight and hearing are
highly developed, as one would expect from an animal which normally hunts
at night in dense vegetation. It is essentially a stalking animal, which with
great skill makes use of every scrap of cover in order to approach
its prey unseen. IS. 244 – 1952

A Turkey Cock painted by Ustad Mansur for the
Mughal Emperor Jahangir in 1612. The Emperor was greatly interested in the
accurate depiction of nature, particularly strange or exotic animals or events. This
turkey was brought from the Portuguese community in Goa to the Mughal court in
1612. Jahangir was fascinated by its strange appearance, especially the colours on its
head and neck, which he said were 'like a chameleon'. It is not known exactly
how or when the Turkey was first introduced to India, but it was brought
from the New World to Germany in 1530. IM. 135 – 1921

OCTOBER

OCTOBER

OCTOBER

In this romantic Kangra painting a lady tries to rescue her pet parakeet
from a young cheetah cub who is making off with it. The cub still has its
white baby fur and wears a typical red collar. IS. 140 – 1955

This highly decorative painting retaining much influence from Persia
shows the Emperor Humayun and Kamran Mirza hunting deer in a rocky mountain
valley. It was painted in about 1800 as a copy of a 17th century original. The
hunting party which is armed with swords and guns and carry
goshawks has disturbed the deer at their watering place. IM. 108 1921

NOVEMBER

NOVEMBER

A group of birds, including orioles, magpie, robin, hoopoes and egrets.
This very decorative 18th century Mughal painting of colourful birds continues a
tradition from the Akbar period in which pages showing imaginary or spectacular
birds were painted against a plain coloured ground. D. 1179 – 1903

Overleaf. Lady with a black buck. This early 19th century Pahari
(Punjab Hills) miniature has clear Mughal influence in the accurate depiction of the
details. The lady, probably a courtesan from the harem, is wearing fashionable fine
gauzy muslin, a cloak beautifully brocaded with an iris pattern and lavish jewellery.
Despite the fact that black buck are frequently used as adjuncts to depictions
of courtly ladies, the black buck, which is here shown tugging at its
halter, was never domesticated. IS. 146 – 1949

In this illustration to the *Akbarnama* Pir Muhammed Khan and
his men drown while trying to cross the treacherous Narmada (Narbarda) river in
1561. Here men can clearly be seen clinging to inflated animal skins. Today, funded
by India and the World Bank, the first of a series of dams has been built.
The local hydrology has been altered to the detriment of some
areas as well as the benefit of others. IS. 2 – 1896 26/117

DECEMBER

DECEMBER

DECEMBER